Everybody Knew

poetry of Bob Komives

Everybody Knew

poetry of Bob Komives

ISBN: 978-1-7338841-2-9

RPK Press
324 East Plum Street
Fort Collins, CO 80524
© 2020

The rules of art and wealth are written in the biosphere, by the biosphere. We are privileged to discover these rules slowly and to use them consciously. In this collection of poems, I look to abundance that has nurtured our biosphere, our species, our art. We blunder when we close our eyes, believing we know we know it all. When we connect ourselves to the truth and mystery in science we help the biosphere and our community create beauty and abundance.

Oil is dirty water
to anyone who has no knowledge of how to use it.
Water is death
to many who do not know how to swim or fish.
Fish may be angels or devils
to people who know no use for them.
Knowing that many fish are edible and tasty
has little value to a hungry quadriplegic
whose arms no longer know how to grasp a fish.
Knowing how to grasp a fish
has little value if it destroys all the fish—
eliminating the knowledge they carry within them.

The biosphere is rich in knowledge of how to organize to convert more and more of the inanimate universe into life. Humankind's wealth—its art and its joys—are in its delicate share of that knowledge.

Bob Komives; Economics of Abundance,
1980-2020

Bob Komives' poetry and life vary in form, tone, and rhythm: serious to silly, personal to theoretical, accepting to judging. In second grade Bob set out to become an architect. After two years with Peace Corps in rural Guatemala and another helping neighborhoods in Little Rock, Arkansas he took his graduate studies and career into Land-use planning. Among his side endeavors he has spent decades exploring and writing about connections between economic theory and natural science. On foot, bicycle, bus, and train he explores landscapes and cultures from the American West to Central America and Europe. Bob loves city and town, farm and wilderness. His mastery of Spanish and continuing study of Hungarian help him find and enjoy depth and breadth in the English language.

Find links to Bob's publications in poetry, art, and economics, as well as to his autobiographical vignettes at:
https://sites.google.com/site/bobkomives/

Thanks to Norman and Gina,
for inspiring me to assemble this "Everybody Knew" selection.
BK

Ruggedless

There are no true rugged individuals.
Either they died
for lack of nurturers beyond birth,
or they were never born
for lack of lovers before conception.

On glare ice flourish
     —if on skates
flounder
     —if in shoes
die there
     —on bare feet

ice or snow
sand or rock
moss or grass
pavement, man-made

oh, groundplain, hold us
     —if well shod—
and blame footwear for each fall

I trim my beard;
you notice my hair.
You cut your hair;
I ask, "since when the beard?"
On sight of a dirty bath,
together we throw out the baby.
My,
what able eyes for change
and lame brains for attribution
have you and I, sir,
and Little Red Ridinghood.

We know.
We know we know the answers.
What to do.
How.
And why.
We know we know.
The false.
The true.
The tried.
Few
are the questions,
and fewer
the answers we cannot grasp.
And at these few
we know to wonder.
Indeed, this is life gifted full:
to know what we know,
yet know how to wonder.
Yet, fuller still in our discovery:
they know,
and they wonder too.
We
and they.
Know
and wonder.
Path to peace!
To human perfection?
No.
This path remains our dream.
And we must know that we know why.
And we must permit ourselves to wonder
how future binds to history.
Today, again:
at which we wonder, they know;
at which we know, they wonder.

We learn from failure, success.
We also unlearn.

Dinosaurs:
      millions of years,
      billions of successes,
      a grand experiment,
      high-class guests,
      then biosphere forgot how to make them.
Humankind:
      new guests,
      still learning,
      biosphere has just begun our experiment.

Among those
species,
cultures,
families,
individuals
who survived yesterday,
many,
but not all,
are fit.
Among those that did not survive,
many,
but not all,
were unfit.
Who will survive until tomorrow?

Countless blunders in the biosphere today.
Some who blundered,
some who innocently stood by,

caught by biospheric justice
—executed before tomorrow.

Let it scare humankind;
again our capacity to blunder rose today.
Let it comfort humankind;
again rose our capacity to understand,
      anticipate,
      record,
      improve
what survives today until tomorrow.

Quandary both obvious and necessary
is not quandary but matter of fact:
River is far more than its valley;
yet valley is more than water within.
Each breeze, more vast than its meadow;
each meadow, greater than its wind.
We find more mystery surrounds us than science
when through science we explore every place.
And child, neighbor, lover, friend:
Our embrace holds far more than meaning;
yet meaning holds all we embrace.

If you have age or wisdom
you have already found
that rare events
eventuate;

Once too rare,
now to common.

Once abundant,
no more around.

This high tide:
noted but briefly,
published for days—hours—
in the line I see etched
after highest ripple ebbs;
to be erased and forgotten
when higher ripples reach their flow.

True high tide:
long remembered,
imprinted for our generation
in the line we see carved
where once highest high wave impressed;
to remain archived and referenced
while tens of thousands highest ripples stop below.

Times change;
Earth and Heaven move
when,
in the marrow of one breastbone,
steep ego
and
deep humility
embrace.

You think Galileo was a great one,
but he wrote heresy in 1632.
He wrote, " Copernicus is right,
our earth circles the sun,
not the other way around. "
You think Galileo was a great one,
but in 1633 he did recant so he would not burn.

Now you too believe
that to the sun belong the planets,
that we live on one example of them,
our sun-centered revolution,
a scientific revelation,
from a genius then among them,
a religious revolution,
insult to god above them.

One way or another,
believers go early, but truth stays late.

Yes, die for your country to get a plaque.
Yes, die for your religion to get guaranteed heaven.
But why die for your science to get guaranteed hell?
Why should you burn for your solar system?
Is the martyr more hero than the genius?
We well know how to make you a martyr,
but we lack the weapon to make you a genius.

How can you resist?
How can you insist:
that Earth is a sphere,

if it is healthier to talk "flat"?
that we came from evolution,
if the inquisition favors special creation?
that all peoples are equal,
if we preach one-ethnic perfection?

Recant today so you will not burn.
Choose humbly to not believe what you believe.
Humility is a sign of greatness.
For everybody knows and the Bible humbly shows
our Earth to be center to the universe.

You think Galileo was a great one
for finding the motion of the pendulum,
the equal rates of falling objects,
and, of course, our telescope.
But then he wrote that Copernicus is right.
You think Galileo was a great one
—but then he did recant
—but then he did not burn.

Once upon a time, everybody knew
the earth is the center of the universe.
This was confirmed by religion
and ratified by politicians.

The best scientists of the day spoke doubts.

Once upon a time, everybody knew
the earth is flat.
Common sense confirmed it.
Common politicians ratified it.

The best scientists of the day spoke doubts.

Since everybody knew,
nobody listened.

I found a faint road through a vast field
where genius, fool, and charlatan must ply.
As hard as the road is to follow,
harder still is to know who am I.

With early victory come sun beams.
Later, come smoke and ash that hide defeat.
We now retreat,
but first we won
over.whelm.ing.ly.
We thought we knew
to administer our place,
to celebrate our art,
to hand our wisdom
to generations that we would save from history.
Soon, we end retreat,
regroup and wait,
to remember how victory came
and forget which way it went.
So, let us post now an order to govern our attack
when we take back the larger part.
Some of us
(duty bound)
must turn coat
to defend retreating enemy
—and there defend our victory.
Early victory brings sun beams.
Total victory portends defeat.
Let us stop this time where the sun shines.

It uplifts the spirit
to walk ourselves back a tiny step toward our planet's nature
to see bright sun
at what only yesterday
was a dark hour.

Our ancestors
(likely even their ancestors)
knew the meaning of sunrise,
of sunset,
and of the high noon between.

Then,
they begat civilization
which begat commerce and industry
which needed to divide the day and night.
Ancestors gave each a dozen hours.

Then,
they begat machines
and the skills to make them,
which begat
a desire to give more precision
to the 4 o'clock meet.
But,
(below and above the equator)
the best machinists had trouble
making their hours shorter
then longer
(and then shorter again)
as our planet's year progressed.

Until,
they added the two-dozens
into twenty-four equal parts,
so the machinists could work their magic,

and the voiceless sun would have to rise
at a different hour and minute each day.
But, it worked—
brilliantly.

Then,
further offspring,
machinists and the mechanics,
invented the steam engine and its railroads.
They made civilization roll into a leap forward
again.

Their descendants,
(our ancestors and their things)
moved so quickly along these roads
that it became a problem
to know the exact hour here
but not there:
"When do you depart?"
"When might I expect you?"
"Can't you just write me down a schedule?"

So, their children,
our ancestors
told us to ignore our personal, local high noon.
They settled quite comfortably into time zones.
Even breathed sighs of relief.
It was good—
for a good time.

Until, their children,
our mothers and fathers,
(at work and at play)
found it hard to give up the summer.
Crazy as it seemed,
little by little,
place by place,
they pushed summer into winter

and called it "savings time."
And most of us say it is good—
for a time.

For,
our ancestors,
our mothers and fathers
in adding more "unnatural"
to the already "unnatural"
gave us a sudden,
pleasant,
yearly,
surprise,
and
(at dark times)
a hope-filled metaphor.

For,
it uplifts the spirit
to walk ourselves back a tiny step toward our planet's nature
to see bright sun
at what only yesterday
was a dark, dark hour.

Ah, Genius of Science,
I confess I owe you much
for chasing off my curable ignorance
and my primitive superstitions.
Yet,
here lingers
  an armchair romance,
    a well-couched prejudice,
      a naive daydream,
        a more truthful confession:
A favorite few came back.

Is this that now hovers
the awaited rain
that will soon descend
to draw our thirst
before refreshment comes to ground?

Or is this no more than past rain that rises
and lingers a while to taunt our thirst
before our fertility departs?

Good Day, Passer-by

Good day, Passer-by,
I watch you pass.
I am pleased when you stop.

You see us all around
as you build your city on our ground.
Perhaps because we do not flee
you think us to be thriving.
But we have lost
ninety-nine percent of our land,
ninety-nine percent of our home ground,
the great North American Prairie
where we build and maintain our towns,
where we built cities
before your species came around.
If we look proud,
we are proud.
Wherever Buffalo roamed we built homes.
We made the great prairie what it was
and can be.
We till the land,
propagate grasses,
release our homes to burrowing owls,
insects,
snakes,
countless species
who do not know how to make a home
unless first we develop the land.
We install utility,
maintain ability
of prairie to be prairie.

Good day, Passer-by.
I am pleased when you stop.
I am Prairie Dog,

a rodent with complex language,
sophisticated society,
commitment to family.
I am your good neighbor.
I know to live with you.
Can you learn and live with me?

I do not care for the eagle,
nor the hawk,
ferret,
coyote,
fox,
nor wolf.
They have taken away family
and want to prey upon me.
Yet I have learned to live with them.
They depend on me.
I am great prairie's great friend.
If you are fond of eagle,
remember me;
I am your great eagle's greater friend.

I am prairie citizen.
On the last one percent of my homeland
I can live with you.
The prairie is passing, Fellow Citizen.
It will slow as you stop,
and it will grow when you show
you can live with me.

I am Fibian,
northern leopard frog.
This place is one of my spots.
I was once so common
along Cache La Poudre River,
up and down
across North America
you did not count me for much.
But such is history.
Today,
you do not count me for many.
I am Fibian,
cold-blooded amphibian.
Relative to many of my relatives
I am fine
(though in precipitous decline)
because still you can find me.
That is hard to say of my cousin,
Boreal Toad,
supposed to live near the mountain headwater.
He is obviously and officially endangered,
while I am just Fibian,
northern leopard frog,
another amphibian
in the official state of special concern.
I am three and one half inches long.
I take well to cold
but do not freeze.
I eat insects
and other things that would upset your stomach.
Some of you say I am spotted green
to be spotted in the meadow during summer
vacation.
Some of you say I am spotted brown
to be spotted in deep water, pond, and wetland.
I crisscross your roads in spring and fall

to prove that none of you is wrong.
But something is wrong
along the river,
around the world.
I am Fibian,
frog, amphibian;
listen up!
When I croak you jump!
In wetland mud I draw a line.
Hop to it and heed my sign:
Beyond this point
river stays open
to song and dance and occasional rhyme,
but valley is closed
to amphibian decline.

Energy begat matter.
Matter begat life.
Life begat knowledge.
Knowledge begat culture.
Then culture begat.

Capture broadly—
  as leaf captures sun,
    mill captures wind,
      and harvester gathers grain.

Distribute deeply—
  as leaf sends oxygen,
    mill delivers flour,
      and parent feeds child
        that teacher educates.

Recirculate densely—
  as we bake for our miller,
    who rewards our harvester,
    who buries our excess
      to reward the roots
        who will feed new leaves.
        to nourish grains of life.

We can, I suppose,
think of a chemical reaction
at the nucleus of a tiny amoeba
as but one reaction
in
a
chain
that
extends
from
this
amoebic nucleus
to the farthest reach of the biosphere.

We should rather, I propose,
notice this amoeba has a surface,
its working edge,
where universe divides in two;
where biosphere divides into
what-is-this-amoeba
.... and ....
what-this-amoeba-is-not

You of leg,
you of twig,
and you, of course, bacterium,
huddle broadly,
densely,
deeply.

Please capture geotherm and passing sun.
Please distribute.
Please recirculate.

For love of wealth
and of biosphere one,
huddle densely,
deeply,
broadly,
you of wing,
you of wheel,
and you, of course, bacterium.

My Common Stove

My stove knows how to burn its firewood,
how to respond
to me who knows so little of what it knows,
to me who does not know how to make a stove.

My stove knows how to send smoke up its chimney
and warmth into my room.
Its warmth can please,
or it can save a half-frozen life.
Such is its success and popularity
that I could sell tickets to my stove's proximity.
But, I do not.
I share its warm knowledge freely
according to communal tradition
among family, neighbors, and kindred strangers.
Those whom my stove knows to please,
those whom my stove knows to save
give back nothing in trade
—except,
to carry forward in common tradition
what we and a stove
must in-common know.

Yesterday, our day that started well,
ended better,
brought success
not perfection
but good enough to compensate for past delay.

Today,
our day to glory
not worry,
but ponder why we chose to worry.

Success came as it had to come
by way of gift
    we did not give
    could not control,
    could not expect,
    but did accept
    from others who assumed success
    and (by their simple gift of faith)
    made success
    all
    but
    in-es-capable.

Tomorrow,
our day to try again
to be again
humble and optimistic.

THE END